History of Syphilis

De Morbo Gallico

Prof. Camillo O. DI CICCO, M.D.
American Association for the History of Medicine
European Academy of Dermatology and Venereology

In A.D. 4 March 1493, Christopher Columbus made return from the "New World", landing to Lisbona from where would have moved, subsequently, towards Spain in order to disembark finally to Barcelona. This is the historical event from which the so-called "americanist theory" draws cue on the origin of the syphilis, that diffused like a terrible epidemic on all Europe towards the end of the '400, accompanying the rising of the Renaissance and the dawn of the Modern Age. Accomplices the movement of armies and populations, for the wars of the time, and blooming of travels and commerce, characterizing the end of the Middle Ages, from the Europe the flagellum scattered very soon to the rest of the world then known diffusing also in North Africa and East and finding the medicine of the time totally unprepared. Ruy Diaz de Isla in the work "*Tractado contra el mal serpentino*", written in 1510 and published in 1539, refers to have cured, during the travel of return in Europe, many members of the shipment of Columbus, affections from certain luetic manifestations and thinks the new disease was imported from Hispaniola (Haiti).

"Tratado llamado fruto de todos los santos, contra el mal serpentino venido de la isla Española, hecho y ordenado en el grande y famoso hospital de todos los Santos de la insigne y muy nombrada ciudad de Lisboa", que fue aprobada por el Emperador Carlos V, por Real Cédula de

10 de julio de 1537. Y que "para no ofender a ninguna nación con el mote denigrativo que cada una se imputaba, nombrándole mal francés, napolitano, portugués, italiano, castellano, tomò un nuevo rumbo".
Ruiz Díaz de Isla (1462-1542).

Bartolomè de Las Casas had conceived the same opinion. In the "*Historia de Las Indias*", wrote that between the Conquistadores the idea of the "bestiality" of the aborigines was prevalent and the disease would have been known already previously in the New World. Moreover the most modern historiography places the accent on the strumentalization of this idea to the aims of the colonial enslavement. The aborigines: lustful, inferiors, "homuncoli" (De Oviedo), naturally needy of being converted and to receive, therefore, with the faith also the slavery. In 1502 Francisco Lopez de Villalobos has published in Venice "*Fumaria de la medicina en romance trovado con untractado sobre las pestiferas Bubas*", which appears to be the oldest book, written in Spanish, on the description of the disease, where for the first time, is used the term: Bubas. Even in Salzburg, the doctor Leonard Schmaus, wrote how the disease is transmitted to the Spanish sailors by aborigines of New World and suggests the care in the book: "*Lucubratiuncola de morbo gallico et cura ejus novita reperta cum ligno Indico*" (1518). In 1527 Jacques de Bethengourt was the first author to speak in France of the new disease, the syphilis, and his venereal origin. He published in Paris:"*Nova penitentialis quadragesima, nec non purgatorium in morbum gallicum, seu venereum; una*

cum dialogo aque argenti et ligni guaiaci colluctantium super dicti morbi curationis prelatur, opus fructiferum".
In this text, the Lent is described allegorically as therapy carried out with the guaiac associated with diet rigid, instead represent the Purgatory, according to the author, excessive salivation caused by the care with the mercurial treatment of syphilis.

In the *"Historia General y Natural de las Indias"* written in 1537, by Gonzalo Fernandez de Oviedo, Governor of the West Indies, is reported the origin American of disease, transmitted to the sailors of Columbus by indigenous women. De Oviedo, who lives, more than a decade in Hispaniola, as an spanish administrator, writes: *"We can take for sure that this disease comes from India and it is was brought to Europe from the New World"*. In his writings De Oviedo says he learned from the natives of Hispaniola, the Taino people, that they were able to cure syphilis using a decoction made by the guaiac tree, which grew in abundance on the island. The disease, endemic and particularly benign in the island, already called by the Spanish, Bubas, term introduced by Lopez de Villalobos, was called by natives of the place, guayanaras, hipas, taybas, isas. Using popular beliefs, and wanting to give a mystical medicine, De Oviedo repeated that where God had sent a sickness to punish sinners, near it, had also created its remedy (lignum vitae – lignum sanctorum). Based on these assumptions, De Oviedo has purchased several estates in the island, where grew in abundance guaiac, and was in partnership with the Fuggers of Augsburg (Bavaria), important family of German bankers and merchants, whoseprominent figure, Jacob Fugger, is still consideredtoday the wealthiest banker of all time. Importing large quantities of wood saint, as was later called the guaiac in Europe, Jacob Fugger managed to increase its enormous luck. The "holy wood" was listed in France, in the 1500, nine livres to a pound (453.5 g.), A figure that corresponded to 25 working days of a specialized worker, such as a French tailor.

This happened while the Taino people, who had over 60,000 attendances in the island of Hispaniola in 1508, because of exploitation, diseases and abuse of the Conquistadores, is later reduced to just 600 units, as reported Friar Bartolomé de Las Casas, in his work:

"Brevisima Relación De La Destruycion De Las Indias".

Theodor de Bry (1528-1598)

The Dominican friar Bartolome de Las Casas wrote to early 1500:

"*All these people of every kind, was created by God without malice and without duplicity [...]. *"

"From History or Brief Relation of distruction of West Indies with its true spanish original already printed by Bartholomew Homes, or Casaus in Seville."

They discovered the Indies the year 1492.
" In the island of Hispaniola they caused, the beginning of the immense slaughter, and distructions of these nations, and which destroyed and deserted;
the Indians, gave them voluntarily, in accordance with the freedom that each had, ... Indians began to realize that these men ought not to be coming from Heaven .... "

"For others, those who wanted and were taken alive, they have cut off both hands and attacked to the body, and said: go and take letters for the people who have fled to the mountains."

"And I saw burn alive, rip and torment with different and new ways of torment, endless people, make slaves all those who, took alive."

"I want to finish, saying, say before God, to my knowledge, that to the Indians were made all the injustice and the evil said, and other which I omit .. .... "

Brevisima Relacion De La Destruycion De Las Indias, Bartolome De Las Casas.
Memorial ordained in 1542 by Emperor Charles V.

The German author Nicholas Poll claimed in the volume
"De cura morbi gallici per lignum guayacanum libellus"
Venetiis, 1535 - Basel, 1536, dedicated to Cardinal
Lange, that in 1517 was able to cure syphilis, from the
wood of the guaiac imported by Hispaniola, describing
how, with a decoction of this wood was able to cure
about 3,000 patients:

"In quibusdam, desperationis causa, nihil medicinarum applicatum fuerat, quorum postea omnium per guayacanum lignum curatio quasi pro miraculo ab omnibus habita fuit: haec enim, uno quasi et eodem tempore, tria fere hominum millia ad bonam valetudinem reducta fuisse, qui post convalescentiam sibi ipsis renasci videbantur".

Even Amerigo Vespucci had shared theory Americanist of syphilis, and stated that women indigenous were highly skilled in terms of sex.

It seems they sting the Spanish sailors, reported
Vespucci, with particular insects, for determine them an uncontrollable sexual desire. The explanation of Vespucci was also reference to theory, much in vogue in this period, that the origin of the disease was determined by lust, violence and rapes no brakes.

Also shares the origin Americanist Pere Pintor, in its
treatise *"De Morbo foedo et occult gallic His temporibus affligente "*. Rome - Eucharius Silber, 1500.

Moreover, we must remember that the very latest
historiography focuses on the exploitation of
this idea for colonial enslavement, to which

were soon subjected the natives:
lustful, morally corrupt, of course, lower
"Homuncoli" (De Oviedo), that need to be converted and
to receive, then, with faith even slavery.
In our view, the denial of diversity and the willingness
of the conquistadores of incorporate the New World
according to the paradigms of their civilization, led
inevitably to a clash with a different culture,
with massacres and unspeakable violence against the
natives, then subjecting the "Aborigines" to the
obedience of the civil and religious laws imposed with the
strength, ie the complete slavery.
Taking possession of the Land of the New World with the
performance, as the first act, of the royal standard, was
experienced by the Spanish as something new, wonderful,
not contradicted by anyone: "Y no me fue contradicho" was
pronounced, landing to the beach of the new
territories, and therefore having full legal effect.

To contrast the theory Americanist, moreover, there is another large group of authors who have not only supported a different theory, called "old" or "pre-Columbian" or "Europeista theory", but they even denied the novelty of disease.

Among the first we want to quote the great historian of medicine Sprengel, which includes the suggestive hypothesis of identification with the syphilis of some forms of disease described in ancient works by Hippocrates, Pliny the Elder, Celsus, Galen, "The ill - Campanian" of Horace, and others.

Exciting, also, the argument of some authors who would see the syphilitic disease already present in the Babylonian Code of Hammurabi (2200 B.C.), where a disease with characteristics similar to syphilis, called Benu, is cause of withholding of contract, in case of sale of a sick slave.

Other cases are described in the Babylonian poem "Gilgamesh", the character of the Mesopotamian myths, mythical king of Sumer, the oldest human settlement in Persia, today's Iraq, near the Persian Gulf.

Also reported other cases of presumed syphilis in the Egyptian papyrus Ebers (1550 BC).

The papyrus, written in hieratic, contains numerous medical indications, with whole chapters devoted to contraception and the prevention of venereal diseases.

Before the year 1500 we find descriptions of the disease in these works:

Leoniceno in Nicholas and Johannes Widmann (1497), Bartolomeo Montagnana Bartholmeus Steber (1498), Gaspare Torella, C. from Pistoia (1500).

In particular, Nicholas Leoniceno, Physician-Humanist known as Leonicenus (1428-1524), professor of medicine at Padua, Bologna and Ferrara, author of prestigious translations of Hippocrates and Galen, wrote in the 1497 one of the first texts on syphilis.

The text "*De Itali epidemic quam morbum gallicum vocant*" suggests an etiology of the natural type of syphilis, sharply in contrast with the thought of Avicenna, that raging in this period, that is to seek the origins of the disease in supernatural phenomena, such the stars, witches, demonic actions, or punishments of religious origin for sinful behavior, "Peccans matter."

Leoniceno, attacking the idea of Avicenna, determined the so-called dispute of Ferrara, who saw the idea of Leonicenus, humanist medicine, opposed to the Arab-Islamic medicine, in a different way of processing to explain the diseases.

The dispute of Ferrara was held like a public debate medical-literary, in the Court of Ercole I d'Este, groped for to clarify the causes of syphilis, its origins and the appropriate therapy.

On the one hand Leoniceno Nicholas, with Sebastian Aquila, making clear what is stated in his treatise, ie that the disease and the resulting epidemic, were determined from natural causes such as famine, floods and pestilence, contrary to what is claimed by a vision Arabic Islamic of supernatural origin.

With a medical concept entirely different, moreover, Montesauro Christmas, professor at the University of Bologna in his treatise:

"De dispositionibus Quas vulgares sore Franzoso Appellant ".

Whatever may have been the result of the dispute, with opposing ideas and variously accepted by the public, the worth of Leoniceno was that from that moment on, the method of establish a dispute in order to compare ideas, studies and research to explain and deal with a disease hitherto unknown, spread from Italy to Leipzig, and

then in France and Spain.
In this way was tried to bring the etiology, diagnosis and
treatment of disease purely on scientific bases,
thus opening the way for new thinking and a
different approach to the disease of medieval medicine.
The so-called dispute of Ferrara in 1500, to discuss of
pathology and therapy, and the similar meetings
of the subsequent years, were the forerunners of the modern
scientific medical congress.

Infected monks (Syphilis?) receive a priest's blessing (Ca. 1360-75 James Palmer's Omne the bonum)

The dilemma about the origin of syphilis, however, is already outlined in "*De Morbo Gallico*" by Giovanni da Vigo (Rapallo 1450 - Rome 1525), who began his studies in Italy using as therapy a red powder, or cinnabar, remembered in the history of medicine as: "*The dust of da Vigo.*"

Closer to the present day, historians of medicine of the authoritativeness of Sudhof and the Italian Castiglioni, have challenged the Americanist theory of the syphilis, with the most varied and compelling arguments.

In the early eighties an anthropologist at the University of Rhode Island, Prof. Marc Kelly, has discovered the skull of a young Indian woman, who died of syphilis, with the nose completely undone by the disease. Since the disintegration process occurs in the course of years, the woman should not had any resistance to the syphilis, and the fact that the Indians were without defenses it would make to think that were the sailors Spaniards to export overseas syphilis. This hypothesis is confirmed by the violence with which were manifested the epidemics of syphilis among the Indians in 1600.

However it should be noted that the most recent studies, molecular biology and biochemistry, conducted on specimens of bones belonging to the Mediterranean Area, dating prior of 1400, have not detected, until today, sure traces of the disease.

Conversely, investigations showed syphilitic lesions in bone remains from the geographical areas of the "New World". In his work, in Latin verse "*De Gonorrea violenta*"

Massimo d'Ascoli Pacific (1400? -1500?), speaking of syphilis and its origin, wrote:
"India me genuit, peperit me Gallia mater,
me alit Parthenopes:
dic mihi quae Patria? ".

Conclusions.
In our opinion it is difficult to clarify in a definitive way
this debate, not this, after all, is our goal.
We just want to suggest that the disease has always existed in an attenuated form, with endemic development and various described, which, however, in a particular historical moment, an accumulation of socio-economic concomitants factors, the move massive of armies and populations for many wars and the flourishing of travels and trades, typical of the modern age, determined an aggressive disease and subsequently a pandemic.
Moreover, in later centuries, though it had not been yet identified the specific therapy and decisive, the disease became endemic, it also at cause of the progressive improvement of socioeconomic conditions and food hygiene.
In this respect, it is useful to remember that even up to the first 900, the only useful drug was only the mercury. It was Paul Ehrlich who in 1910 introduced the arsenobenzole
in the treatment of syphilis and, subsequently, in 1928 Fleming discovered penicillin, however, produced on industrial scale in the 1941.

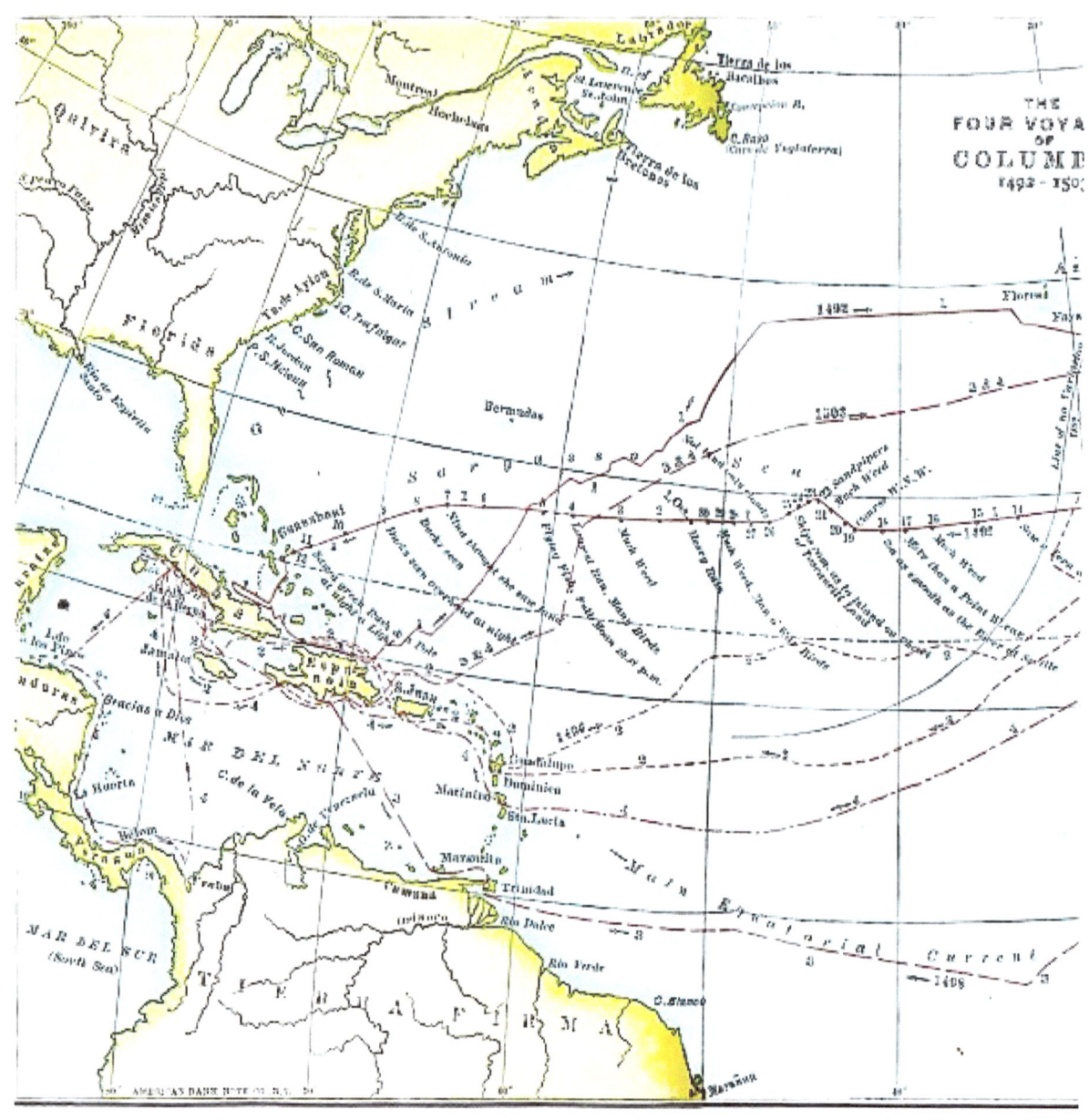

The four travels of Columbus (1492-1503)

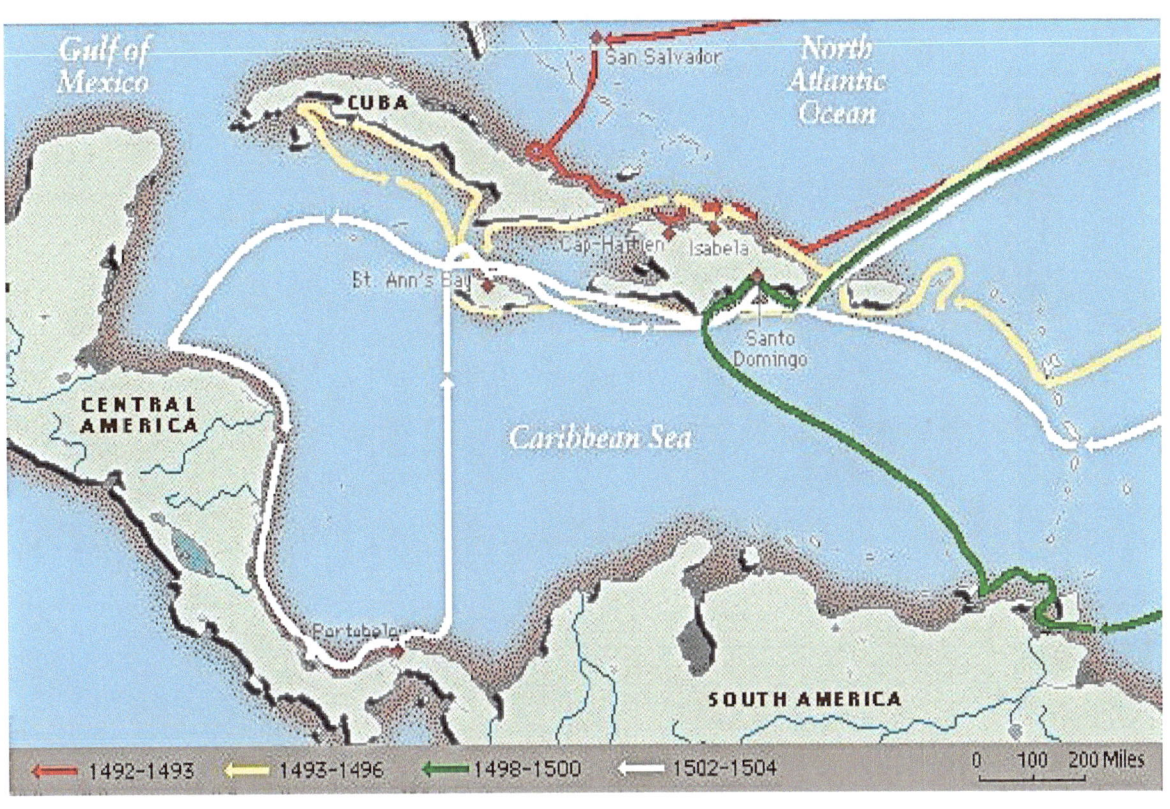

Photoreproduction from Theodor de Bry and Charles de la Roncière, *La Floride Française: Scènes de la vie Indiennes, peintes en 1564*
[facsimile of the 1564 original (Paris, 1928)].

THE SYPHILIS IN ITALY

In Italy the disease was manifested in epidemic shape in 1494 with the siege of Naples by the French troops on the orders of Charles VIII who died to the age of 28 years, possibly from cerebral syphilis (G. Del Guerra).

A group of approximately 800 prostitutes was aggregated to the French troops and not there is doubt that just the dissemination of the prostitutes in the armies and between the population contributed in maximum part to disseminate the syphilis that in Italy was called "Gallica Plague" or "mal francese", while for the French it was "mal napolitain" or "petit verole", for the Germans "Frantzofen", for the Spanish "Las Bubas", for the British "Grand Gor" or "The French Poches".

Pazzini writes in *"History of diseases"*: Invaders armies had alongside another army of females and who therefore spread and propagated the contagion, both within the ranks of soldiers that among the civilian population".

In the frantic search for a guilty, the people's fantasy, pointed to the origin of the disease as a result of intercourse between a prostitute and a leper.

(Paracelsus, astrologer, alchemist, physician, 1493-1541).

Also guilty of spreading the disease were indicated the Marranos, or converted Jews, or the Indians, the sailors from the New World etc....

In Rome, towards the end of the '400, clandestines excluding, were available approximately 6800 prostitutes. Campo de Fiori was the central point where the prostitution was praticed, and in a prayer to Holy Father to remedy it, we read: " *In this city, the prostitutes are walking stiffly and proud on the streets as matrons*". In Venice the prostitutes were forced to walk with a yellow handkerchief around the neck like sign of acknowledgment.

Sexual abstinence was the measure that the Church adopted as a remedy in order to avoid such disease and Pope Paul IV, around to the half of the '500, decreed with an edict an evicting from Rome and all the Papal State of the prostitutes.

The popular rebellion forced the Church to find a place where to practice the prostitution, away from the city and was decided for a location across the Tevere: today Trastevere.

In the "*De preservatione a carie gallica*" of 1555, Gabriele Falloppia had devised one individual protection against syphilis consisting in one patch of linen to shape of bag, "ad mensuram glandis", soaked with mercury: it was the forerunner of the modern condom.

The continued spread of the disease led to adopt further restrictive measures, tolerating only the opening of certain "houses" where prostitution is exercised, from which the derivative as "houses of tolerance".

At the entrance of these brothels reigned written:
"*Non proestat foemina morbo infesta*".

In Paris, women found to be syphilitic, historians recount,

were isolated and even flogged, before and after the
care. In Vienna, to the common people, was allowed entry,
into the treatment sites, for laugh of them.

The disease began in the genitals with vesicles similar to
the smallpox, therefore was also called Spanish Smallpox.
By Antonio Benivieni (1443-1502) who describes the
evolution of the disease:

*"primo natus morbus incipit a pustolis, hae fere semper in
partibus genitalibus, raro in capite, sed quidam per totum
corpus apparebant, hae apparent variolis simile."*

Antonio Benivieni was a florentine physician, pioneer of
autopsy, which allowed to begin to understand the
causes of death. Benivieni published a treatise "*De Abditis
Morborum causis*" considered one of the first scientific
works of human pathology in which he describes the first
case of congenital syphilis.

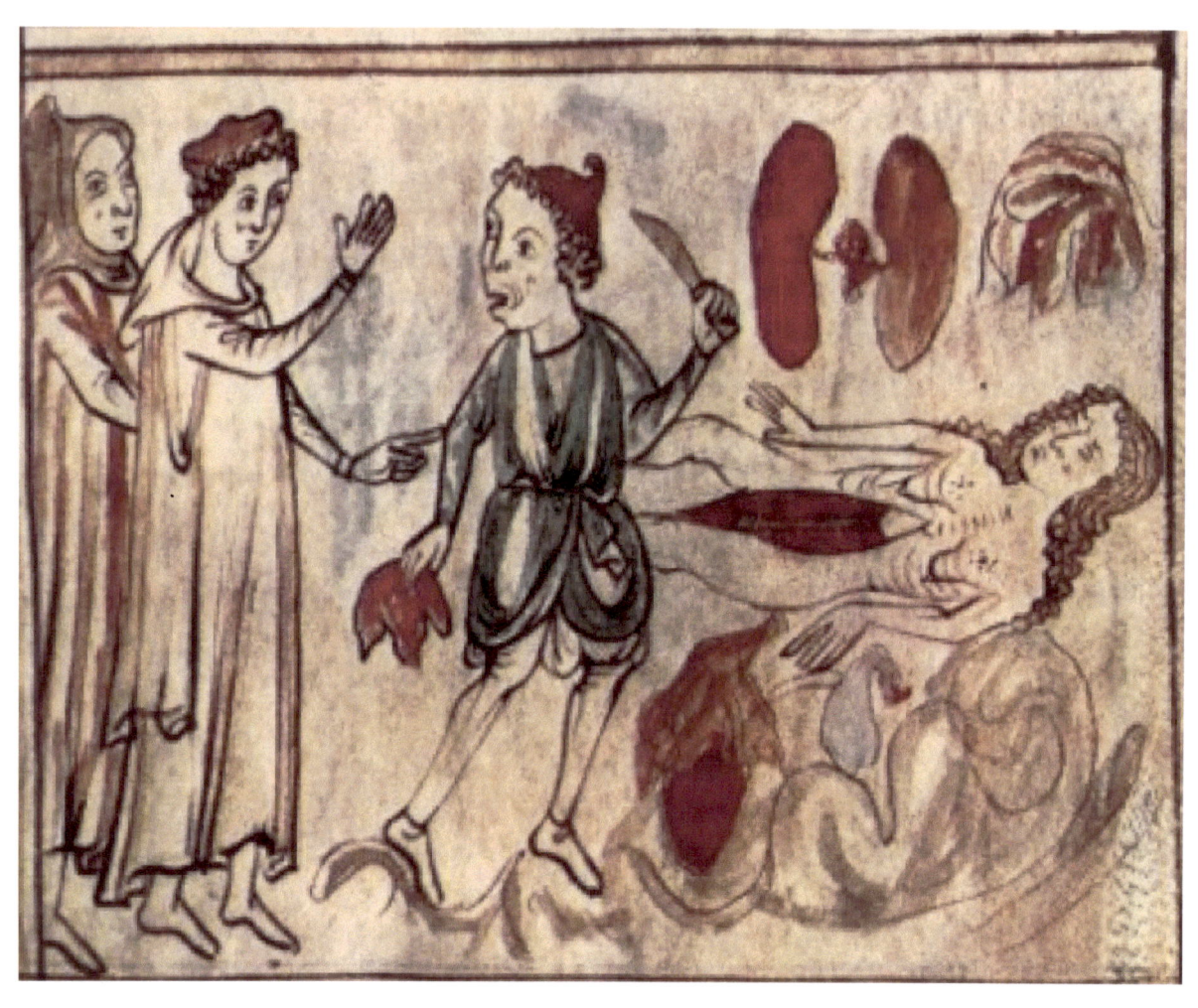

The post-mortem anatomical studies were very rare during the Middle Ages, especially for religious reasons. This painting (c. 1300) shows a surgeon and a monk during an autopsy.

In 1540, Niccolò Massa (Venice 1489-Venice 1569),
author of one of the first books of anatomy (*Anatomiae Books Introductorius,* 1536), describes in detail the various stages of syphilitic disease, indicating that the syphilitic ulcer appeared to be "*rebelia ad omnia rimedia*".
Significant contribution to medical science as the author of works on anatomy, embryology and physiology,
certainly in this period is Hieronymi fabricii of Acquapendente (1553-Padua, 1619).
Physician of great fame, we find him in a letter of
Galileo Galilei (1604), where he is recommended to Christine Lorraine, wife of Francis I, for the healing of his son, Don Carlo de 'Medici, who suffers from a serious illness.
Erasmus of Rotterdam in the work "*Adolescentis et scorti* ", defines the prostitutes "public sewers" and cause of
spread of syphilis.
For those who do not known Latin, heritage
exclusive of the upper classes, were written poems in vulgar, more understood by the people, who described the sad end of the prostitutes affected by Syphilis:
"The pride of the Ferrarese courtesan"
"The Purgatory of courtesans"
"Lament for be reduced in the cart for the syphilis"
"Ballade de la large vérole" Jean Droyn, Lyon 1512.
Works such as *"The Lament of Strascino"* by Nicholas Campani, were very popular and were sung and
recited in the villages squares.
Finally, we recall an important work, with skilfull

description of the syphilis, little known, written only in vulgar Italian, so that can be disseminated and understood by the lower classes:

"*The Enarratio Satyrica*" written by Giorgio Sommariva in the years 1494 – 1495. Sommariva had used the mercury in the treatment of syphilis.

Jerome Capivaccio, (or Capodivacca), born in Padua in early decades of the sixteenth century and died in Mantua in 1589, in his "*De lue venerea acroaseis*", published in 1590, explains the utility of the use of decoctions of guaiac or mercury for fumigations in more severe cases, while believed poorly effective the antimonium as therapy of the syphilis. Patients with syphilis, as happened previously with sick of the plague and leprosy, were removed from the normal places of care and were confined by the authorities in the special hospitalizations.

In Paris in 1557, at the Petites Maisons, in the same site where once stood the lazaretto Saint Germain, were confined syphilitic patients, and mentally ill patients, away from the society. In this period, the Brothers of Divine Love began to cure the syphilis by building the so-called: Hospitals of the Incurables.

Nevertheless, the disease continued to spread and kill people in all walks of life, without sparing clergy and nobility. Distinguished patients were Pope Alexander VI (Rodrigo Borgia), King Charles VIII of France, Albrecht Duhrer, Henry VIII, King of England, Queen Catherine of Aragon, first wife of Henry VIII, King Francis I of France, Pope Julius II and Queen Elizabeth I.

Painters such as Edouard Manet and Paul Gauguin. Composers and writers such as Hugo Wolf, Frederick Delius, Scott Joplin, Gaetano Donizetti, Charles Baudelaire. Cynically, Erasmus of Rotterdam stated: "*A man noble, without syphilis, was not too noble or too man.*"

The first poet who sang and died of syphilis was Antonio Camels, who contracted the disease in 1494. Celebrated in life and throughout the sixteenth century, Antonio Camels, said Pistoia (born in Pistoia in 1436 and died in Ferrara, April 29 1502) was one of the greatest poets of italian burlesque. The Camels is a surname derived by an intimidation to appear to the Lords of Pistoia in 1479, for a dispute of debt.

"*Of all the things that you see, makes sonnets*" was his emblem and quoting the poet Ovid, from the work "Tristia", "*quod tentabam dicere, versus erat*".

The Pistoia, speaking of his syphilis, which led him to death in 1502, used to say so in his burlesque sonnets: "*de novo elected between Franza' baron*".

The religious make appeal to the protecting of Saint Giobbe, Saint Dionigi and San Lazzaro, the patron saints of syphilis.

The disease was understood as a divine punishment and in "*Cicalamenti del Grappa,*" by Anonymous, it is argued that turns out to be good for humanity, because it serves to the purification of the sinner and to the improvement of costumes. Naturally the description of the disease occurred in way superficially grotesque and for hiding the suffering physical and mental of the sicks with syphilis, devoid of appropriate therapy.

Astrologers in this period studied the planets looking
the remedy to the negative conjunction of Jupiter with
Saturn in the sign of Scorpio, a harbinger of disaster.
Their reference point was the astrological handwheel
against syphilis by the doctor Theodoric Ulsenio
Nuremberg, 1484.
The same Ulsenio Theodoric wrote a poem in 1496
"*De Pharmacendi comprovata ratione*" cosidered
an important example of a scientific essay in verses.
According to the theory Hippocratic, that with fusion
syncretic to astrological principles survives until the
Renaissance, is precisely the imbalance between the
primary elements that causes the
disease, both as influence on the individual, either as
epidemic, causing the destruction of the masses.
And just imagine that this happening when it is
seen the massed planets in the sign of Scorpio
(Water sign) in the sky in 1484.

The image of a man plagued by the "Gallic disease", is designed by Albrecht Durer as an illustration of medical prophecy of Theodoric Ulsenio from Nuremberg, and is considered the first illustration that has been printed on syphilis (1496).

The Arrangement of the Universe at the time of Creation, from *Tractatus de pestilentiali scorra sive mala de francais* (1496) by Joseph Grumpeck

Albrecht Durer

Albrecht Dürer (self-portrait of 1500)
(Nuremberg, May 21, 1471 - Nuremberg, April 6, 1528)
Latin inscription: "I Albrecht Dürer of Nuremberg, at the age of twenty-eight years, I have created me with eternal colors the same in my image"

Pope Alexander VI (Rodrigo Borgia)

HISTORY OF SYPHILIS THERAPY

Many treaties were published, trying to find a appropriate therapy to treat the syphilis, and also trying useful advices in order to eat and drink, to prevent the disease. Konrad Schellig in 1496 wrote: *"In pustolas malas morbum quem malum de Francia vulgus appellat que sunt de genere formicarum salubre consilium"*.
Torrella Gaspar publishes the *"Tractatus cum consiliis contra pudendagram seu morbum gallicum"*, Roma, 1497.
Widmann Johann publishes in 1497 *"Tractatus de pustulis et morbo qui vulgato nomine mal de franzos appellatur"*.
But, Oviedo y Valdes wrote in *"De relación sumaria the natural historia de indias* "Toledo 1526:
"..... Just as God's mercy allows us to be afflicted for our sins, so it creates a remedy for our pains".
Obviously Oviedo was referring to the effectiveness of lignum vitae, which is well soon spread throughout the West and praised for its significant therapeutic abilities in many diseases, so called, tree of life, or holy wood, Ligno Holies. The idea spread throughout the populace of the therapeutic efficacy of guaiac and the importance for the mystical aspect in the treatment of syphilis for the Church, had meant that Cardinal Bembo required to Fracastoro to write a third part of his famous *"Syphilidis sive de Morbo Gallico"*, where was described only the treatment of syphilis with the wooden saint, "Ligno Holies".

EL LIGNO SANTO DE INDIA
(The guaiac)

Miraculous healing powers were therefore attributed to
guaiac wood of the Antilles, called "Holy Wood", and
the spanish priest Francisco Delicado (1480-1535), sick of
a serious form of syphilis, explained the use of guaiac
in the treatment of syphilis in his manuscript "*El modo de
adoperare el legno de India* "(1525 Rome).
In the work "*Lozana Andalusia*" the same author described
also as a spanish courtesan in the city of Rome,
made use of substances useful in preventing syphilis and
other venereal diseases.
The work, published in Venice in 1528, narrates the erotic
adventures of an Andalusian courtesan in
Rome in search of fortune during 1524.
In the text the author writes:
" *Whose will come after these punishments, contemplated
this portrait of Rome, and that nobody of you has to inspire
another! Consider his die, the vengeance of heaven and
land, because the elements of the earth rebelled, the people
went against the people, the stranger became oppressive,
and then it was the earthquake, famine, pestilence, flood,
we saw the water not only pursue, but inflate the Tiber to
the point that overflowed and flooded the town, so that on
January 12, 1528 reached the signal of the 1515.
No one escapes to the divine providence, and if with the
guilty will stop the pains, will not stop the sun, moon and
the stars to predict the reward that each will have. "*

Experts physicians recognized in the use of guaiac were:
Paulus Ricius of Habsburg, Heinrich Stromer (1482-1542)
professor of pathology Leibnitz, Gregor Kopp of Kalbe
(Saxony) personal physician of Emperor.

According Huldric von Hutten, with just a decoction of
a pound of lignum vitae, a treatment for four days
in an overheated room and a strict diet, was sufficient to get
excellent results in the treatment of syphilis.

Hutten with the text "*De Guaiaci Medicine*" became an
active promoter of the interests of the banker Fugger
(Family Fugger banking), which held the monopoly on the
sale of guaiac throughout the West.

Unfortunately, even the death of Hutten, in 1523, happened
because of the syphilis. The guaiac was imported to Europe
from India in 1508 and its use in the treatment of syphilis
began in Italy in 1517.

The Guaiac is a slow growing tree with small leaves,
lanceolate, with blue flowers, which bloom in the spring.
Produces a yellow fruit and the wood is hard and
heavy.

The decoction made to treat the disease was hired to
30 days, keeping the patient in the diet and in environment
superheated to foster a deep sweat, considered beneficial.
Unfortunately the decoction called "Holy" was virtually
ineffective in the treatment of syphilis.

Il Guaiacum sanctum is currently the national tree of
Bahamas.

In Rome, at the Hospital San Giacomo for the Incurables,
the Guaiac, considering the cost, was distributed free to
people with syphilis.

Doctors of the time visited the patients twice a day, observing their urine and advising them so what to eat or drink for the treatment of disease.

Flocked to St. Giacomo Hospital hundreds of people with syphilis from each part and in a book of acceptance, conserved in the historical archives of S. Spirit, in the year 1525 are marked some two thousand patients for the treatment of this disease.

The hospital began as an act of expiation of the family Colonna, to repair the slap of Sciarra Colonna to the pope Bonifacio VIII in the conflict of Anagni in 1302.

The Hospital was constructed by Cardinal Pietro Colonna, who died in 1326, cured the sick pilgrims, poor orphans, repented, exiles, old and needy children "proietti", ie the sons of no one.

At the St. Giacomo, first to treat
an ulcer, subsequently devoting himself to the sick
for four years, we find S. Camillo de Lellis, who in 1584
founded "The Fellowship of Ministers of the Sick"
approved by Sisto V in 1586 and in 1591 by Pope
Gregory XIV as an Order religious. His teacher
was S. Filippo Neri, in 1886 S. Camillo is recognized
by the Church, under the papacy of Leone XIII, patron of
patients and hospitals.

In large letters on the front door is written:
*"If you can heal, heal; if you can't heal, calm; if
you can not calm, consoles"* Dr. Augusto Murri.

The high level of the hospitals of Rome in this period is documented by what wrote Martin Luther, a guest of the

Augustinian convent (now the barracks of the Carabinieri in Piazza del Popolo):

"In Italy the hospitals are equipped with everything that is necessary, they are well built, eats and drinks well and will was served promptly, the doctors are good, the beds and furniture are clean and well kept .... veiled ladies are to guard the sick. "

In 1532 Benvenuto Cellini, who was hit by syphilis,
" I received the syphilis by a beautiful young servant " as wrote, began the treatment with guaiac, with the confinement in a locked room and heated for over a month, with a strict diet, and regular daily purges, with daily ingestion of decoction of guaiacum.

The scarcity of therapies for the treatment of syphilis was definitely the cause of the rising cost of guaiac.

Inevitably, the interest in the sale of guaiac, which would derive huge profits, was remarkable by banks and merchants of the time.

The management of the sale of guaiac finally became Family Fugger banking monopoly.

Therapy with Guaiac

Assistance to a person with syphilis.
Engraving of John Stradano (XVIII century)

BANKS AND SYPHILIS.

Jakob Fugger II (1459-1525) and his family,
possessed mines of silver, gold and copper, moreover,
spice and wool, and silk factories operating in Asia.
Known as "Fugger the Rich," coined money and
possessed banks in every European capital.
Fugger, owner of the contract to manage the money of the
Papacy, through its subsidiaries, collected money for the
remission of sins, the sale of indulgences and
the obtaining of ecclesiastical benefices.
Jakob Fugger financed the election of Charles V to Holy
Roman Emperor, which was obtained with bribes of money
corresponding to three times the annual turnover of
Florence (850,000 guilders). While, for Francis I of France,
the amount allocated by the Fugger was 543,000 florins.
Jakob Fugger was harshly criticized by his contemporaries,
especially by Martin Luther, especially for the revenues
from sale of indulgences and of ecclesiastical benefices.
Fugger used a own fortune teller for predict and manage the
results of his numerous affairs. He died in 1525, leaving
two million guilders and beyond seven tons of gold.
Jakob Fugger is still considered one of the people
richest of all time.
He liked to remind his interlocutors:
"The king reigns, but the Bank rules."
Even today there is a settlement where there is a prayer
for Mr. Fugger and his descendants, so they would join in
"Pearly Gates" (The Gates of Paradise).

Praying for the Fugger is an indispensable condition for living in the village, at an annual rent of 1 Rhein guilder, the same from 500 years.

Cash today, 88 cents, or about $ 1.23.

The citizens promise to say three prayers - the Our Father, Hail Mary and the Apostles' Creed - every day, for the souls of the Fugger family of bankers, increasing thus their celestial ambitions.

For this reason at Fuggerei, picturesque village of southern Germany (Augsburg, Bavaria), the rent is not been changed since 1520.

After 1525, with the death of Jakob Fugger, even when the Fugger Family Banking no longer had the monopoly of Guaiac, the price was high, since this therapy was very coveted.

Recall that in France the cost was nine livres for a pound (453.5 g.), a figure that corresponded to 25 working days of a tailor.

MERCURY THERAPY.

Herman Boerhaave (1668-1738) dutch physician and chemist and botanist, considered the founder of the clinical teaching and of the modern hospital wrote: "*Jacobus Berengarius carpensis inunctionis ex .... hydrargyro treated luis venerete primarily fuit inventor* " Berengario da Carpi (1466-1530), known as Jacopo da Carpi, or under the pseudonym James Barigazzi, graduated in Medicine and Philosophy 4 August 1489 to the University of Bologna. He was a physician-surgeon and professor of anatomy in Bologna and Pavia.
Expert in both ancient and modern medicine, what made him an eminent teacher than his contemporaries, such as Achillini (1463-1512) and Zerbi (1445-1505).
Anatomist, published in 1514 an edition of "*Anothomia*" Mondino of Liuzzi (1316), then in 1518 published the "*Tractatus de fractura calvae sive cranei* ".
Berengario da Carpi, used the mercury in the treatment of syphilis, associating to this purpose, the use of a new antiluetic drug (the so-called Sanctum Ligno, "holy wood / tree" or "Guaiacum") and proposed, this therapy in a new edition, August 4 1489, of the little treatise, "*De guaiaci medicina et morbo gallico*".
Huldric von Hutten for nine years suffered the side effects of mercury therapy to no avail. Therefore started treatment with guaiac and described this treatment in the book "*De guaiaci Medicine et Morbo Gallico liber unus*" (1518), dedicated to the bishop Albrecht of Brandeburg, also died

of syphilis. Among the various hypotheses and suggestive therapeutic remedies, the mercury was the only drug to have a even minimal effectiveness.

The Arab physicians were the first to use the mercury locally, in the form of an ointment. In 1025 the Persian physician Abu Ali al- Hussein, known as Avicenna (980-1037), in his famous book *"Qanun fit-Tibb at"* translated into Latin by Gerardo da Cremona as *"Liber canonis medicinae "*, suggests the use of mercury in syphilis.

Successively, this therapy came in Europe and beyond for use in the form of an ointment / poultice, mercury was widely used in fumigation.

The patient was placed in a dark room and heated, was covered with ointment of mercury and placed in a barrel, clear head and feet placed into a brazier that burned resins, gums, cinapro.

In the days after, it was obvious to note stomatitis and intense salivation, which was understood as eliminating of the disease by the patient, but that it was determined solely by mercury intoxication.

The mercury was causing so many deaths from poisoning and, being a poison, also determined numerous side effects, and an improvement of the disease.

Therefore, appealed the sentence at this time

"A night with Venus, a lifetime with Mercury".

THERAPY OF SYPHILIS WITH THE
MERCURY

DIFFERENT METHODS USED IN THE TREATMENT OF
SYPHILIS WITH THE MERCURY

Gerard or Gérard de Lairesse
(11 September 1640 or 1641 - June 1711)
Deformation of the face from syphilis. Rembrant.

THERAPY OF SYPHILIS

In the first half of the twentieth century, medical science transformed the syphilis, a disease that sowed death and hopelessness, in a treatable and curable disease.

In 1906 Fritz Schaudin identified the spirochete Treponema pallidum cause of syphilis and in the same year was introduced for the diagnosis of syphilis, the "Wassermann test". In 1908, the first antibiotic to be used in therapy of luetic infection was "Salvarsan" (a treatment containing arsenic) by Sahachiro Hata, while he worked to the laboratory of Paul Ehrlich, who was later amended in Neosalvarsan.

In 1928 Alexander Fleming began the definitive therapy for eradicate syphilis with the discovery of the inhibition of bacterial growth, due to the action of a substance secreted by the mold: the "Penicillin".

In 1928 happened that Fleming had been absent from his laboratory for a short break, while was working on some strains of bacteria, grown in a culture dish.

Occasionally, after returning from holiday, Fleming noticed in a capsule the presence of a clear halo unusual, where the Bacteria weren't grown.

In the center, it was possible to note the presence of a mold that had contaminated the crops and therefore Fleming concluded that it was the cause of death of bacteria.

This phenomenon is reproduced by cultivating in the same soil the mold Penicillium notatum and various strains Bacterial. In this way, was observed around the colony of Penicillium, an empty region, in which the bacteria do not

were able to grow, so the bacterial colonies
there were only confined to the outer end of the
culture medium.

"The story of penicillin has something of romance and helps to explain the weight of fate, luck, fate or destiny, as you want to call in career of each person. "
Alexander Fleming.

Duty in the history of therapy to remember that in 1895 the doctor Vincenzo Saverio, born in Sepino, Italy, physician of the Naval Medical Corps, published the results of his research, where showed the power bactericidal of the molds of the genus Penicillium on various bacteria (Staphylococci, streptococci, typhoid bacilli, Vibrio Cholera etc..), concluding that "*the properties of these molds are a major obstacle for life and propagation of pathogenic bacteria. "*
The intuition of the doctor sprang by the observation of a well, where was found the presence
of drinking water, only in the presence of a substance gelatinous, green color, adhering to the walls of the well.
Conversely, the water was being polluted in his absence and cause of gastroenteritis. This framework was also
known by local farmers for use of the water wells.
The dott. Vincenzo Saverio hypothesized, therefore, the presence of a substance created in the green gelatinous material, which prevented the proliferation of germs cause of infections.
Subsequently identified in the mold of the Penicillium the bactericidal power, by publishing his studies in the

47

"Annals of Experimental Hygiene, year 1895", University of Naples, with the title:

"Sauces extracts of certain molds. Researches of Dott. Vincenzo Tiberio ".

The report published in the journal Annals, only in italian language, unfortunately, remained locked up in italian laboratories and remained totally unknown abroad. The doctor died of a heart attack while he was studying the mold Penicillium notatum and various strains Bacterial. All this happened 33 years before the occasional discovery of penicillin by Alexander Fleming.

In 1940 two researchers at Oxford, Howard Walter Florey, and the German, Ernst Boris Chain, both Nobel along with Alexander Fleming, using technical advances, improved the insulation of Penicillin and its concentration. Was used in this way, the penicillin to treat effectively not only the animal of laboratory, but also the man, what that happened Feb. 12, 1941, with the healing of a patient suffering from a severe form of septicemia, in just 24 hours. However, the price of penicillin was still high and, despite his efforts, the amount produced was still poor and certainly not usable on a large scale. Fleming thought of enriching the land where they were made grow molds producing penicillin and was obtained in this way a ten-fold increase in the production of antibiotic.

It is also thought to flow from around the world, different samples of mold Penicillium, with the assumption that some varieties were able to produce a lot of penicillin unlike other, in order to identify the type of mold capable of producing the greatest amount of antibiotic.

Occasionally a woman, Mary Hunt, discovered on surface of a melon that she have purchased, a pleasant and colored mold that caused her to bring a sample to the laboratories of his town, Peoria.

The mold posted by Mary Hunt, was considered optimal in the production of significant amounts of penicillin. Therefore was selected as the best and then was called: *mold of Mary.*

This fact increased by tenfold the production capacity, and started in this way the mass production of penicillin on a large scale, that allowed to save the life of millions of people around the world. The "sterilisatio magna", advocated by the German bacteriologist Ehrlich, therefore was achieved with the introduction of the penicillin in the medical therapy: in this way began Antibiotic Era.

TUSKEGEE EXPERIMENT.

In the history of syphilis, appears to be important bring to knowledge this experiment conducted on the disease,
came to light after many decades, so as to not repeat it, in view of clinical research and study of other diseases, present or future, that still we do not know totally, such as AIDS.
On July 25, 1972, the "Washington Evening Star" published an front-page article entitled "Syphilis Patients Died Untreated". With these words the America came to knowledge, after 40 years, of a study in patients with syphilis, untreated and studied until death, on order to observe subsequently, by autopsy, such as the organs were damaged.
The day following its publication all the U.S. newspapers reported the news of the Tuskegee experiment, from the name of the Tuskegee Institute a prestigious African-American College of Alabama, who, in collaboration with the U.S. Public Health Service, began the study in 1932 enlisting for this purpose, by the County in Macon Alabama, 400 black and poor people, suffering from syphilis.
The patients, as reported by the Centers for Disease Control, CDC, were not aware of being infected with syphilis, was reported their, instead, be a treatment for "Bad blood", a term used to describe many diseases.
Although in the years after the penicillin was widely used in the treatment of syphilis, the Tuskegee experiment

continued in the same way until 1972, when was interrupted by the spread of the news, until then remained secret.

As reported by the *New York Times* on 25 July 1972, the Tuskegee Syphilis Study, was revealed as "*the longest nontherapeutic experiment on human beings in medical history*".

In 1979, was wrote the "Belmont Report", an important historical document in the field of medical ethics, created by "Department of Health and Human Services U.S.A.", document entitled Ethical Principles and

Guidelines for the Protection of Human Subjects of Research.

The report was written April 18, 1979 and owes its name to the Belmont Conference Center, where was processed this document.

Only 65 years later, May 16, 1997, the government has asked formally apologize to the victims for the study immoral.

So President Clinton: *"For the survivors, their wives and family, children and grandchildren, I say: no power on Terra is able to restore the lives lost, the pain suffered, the years of internal torment and anguish. What was done can not be undone. But we can put an end to the silence. We can stop turning the head far. We can look in the eye and finally say, behalf of the American people: ... What has been done by U.S. government is shameful. And I'm sorry"*.

While the "Tuskegee experiment" was proceeding by studying the syphilis in patients already sick, was effected in Guatemala another study with Treponema pallidum infecting healthy people.

Prostitutes infected with syphilis were paid to infect inmates of prisons in Guatemala, in other cases, the researchers carried infection directly.

In the period 1946-48 the federal investigators of the Ministry of Health of the United States, under the guidance of Dr. John Cutler, have deliberately infected 1,308 people in Guatemala with venereal diseases and subsequently have studied them and their children until 1953.

The study was funded by the National Institutes of Health and by the Pan American Sanitary Bureau, with the approval of Guatemalan Minsters, who have been engaged to preserve the confidentiality of U.S. operations.

This study came to light in the year 2005 thanks to the studies performed by a medical historian, Prof. Susan Mokotoff Reverby, Wellesley College (Maryland), studying the records of Dr. John Cutler, the former head of the Tuskegee experiment.

Involving poor people of color in a study on syphilis, whose purpose was only to observe the progression of disease without performing any treatment, patients were enrolled from psychiatric hospitals, prostitutes, soldiers, prisoners, exposing them to syphilis, gonorrhea, chancroid. The researchers carried the study, being to knowledge that in this way were violated fundamental rules of ethics at the base of each search scientific. The thesis of the researchers, as reported by the U.S. Bioethics Commission, was to assess the effects of penicillin in infections of the venereal diseases, treating only 700 patients and leaving the other without treatment to assess the progress of the disease. Subsequently the results showed that these studies were immoral, racist, without any additional medical knowledge bringing to light some experiments of exceptional gravity, documenting attitudes unimaginable, that should never be applied to medical research.

Already in October 2010, the U.S. government apologized and formally announced that for the violations of Rights Human committed during these experiments, will not be worth the Prescription of offenses.

On August 30, 2011, The Guardian publishes sensational revelations about the experiments carried out by Researchers in Guatemala. At the "'Asylum de Alienados", so called the home for the mentally ill, was made an injection at the base of the skull of Treponema pallidum (syphilis) in seven women with epilepsy, wanting to experiment so a new treatment of the disease, but determining only a bacterial meningitis. Without principles of ethics and morality, was also create an additional infection in a patient with end-stage syphilis, infecting her eyes with gonorrhea in order to evaluate the consequences. Six months after the patient died.

Dr. Amy Gutmann, head of the commission, described the case as "egregious coldly," forgetting that the experiment will be remembered by the history as a shame for medical science.

President Barack Obama has apologized to the President of Guatemala, Alvaro Colom, and also ordered that the Bioethics Commission should proceed to analyze again the experiments effected in Guatemala. The Obama administration also have announced the appropriation of one million dollars to study new rules to protect research volunteers medical, and a further sum of $ 775,000 for the combating sexually transmitted diseases in Guatemala, as reported by the Washington Post. The U.S. Bioethics Commission, considered the fact, came to the conclusion that what happened turned out to be an absolute brutality. A comparison was made possible with the experiment done by Cutler, in 1943, studying the gonorrhea on infected detainees in Terre Haute. In this case, however, it should be noted that the detainees had been made aware of the study and thus, they had given their consent to testing. Instead, in many cases, the participants of the experiment Guatemala were not aware of the study and not gave their consent to testing. In a joint statement, Barack Obama, Hillary Clinton and Katheleen Sebelius (Secretary of Health and Human Services of the United States) have stated:

"Despite these events allegedly occurred more than 64 years ago, We are outraged that such reprehensible research has been conducted under the guise of public health. We deeply regret that this happened, and we apologize to all those who have been affected by these aberrant searches. The behavior shown in these researchs does not represent the values of the United States, or our efforts to promote human dignity and thegreat respect for the people of Guatemala".

The New York Times

Syphilis Victims in U.S. Study Went Untreated for 40 Years

By JEAN HELLER
The Associated Press

WASHINGTON, July 25—For 40 years the United States Public Health Service has conducted a study in which human beings with syphilis, who were induced to serve as guinea pigs, have gone without medical treatment for the disease and a few have died of its late effects, even though an effective therapy was eventually discovered.

The study was conducted to determine from autopsies what the disease does to the human body.

Officials of the health service who initiated the experiment have long since retired. Current officials, who say they have serious doubts about the morality of the study, also say that it is too late to treat the syphilis in any surviving participants.

Doctors in the service say they are now rendering whatever other medical services they can give to the survivors while the study of the disease's effects continues.

Dr. Merlin K. DuVal, Assistant Secretary of Health, Education and Welfare for Health and Scientific Affairs, expressed shock on learning of the study. He said that he was making an immediate investigation.

The experiment, called the Tuskegee Study, began in 1932 with about 600 black men,

FRACASTORO, Girolamo (1478-1553)

Girolamo Fracastoro (in the literary bibliography known also with the Latin name Hieronimus Fracastorius), was born in 1478 to Verona, in that time still Republic of Venice, from a noble family.

He studied at Padova University, where he was graduated in 1502. At the same University, was assigned him the chair of Logic and Philosophy.

His teacher, was the doctor-philosopher, Pietro Pomponazzi, instead, his study colleagues were: Andrea Navagero, studious and historian, Pietro Bembo and Gaspare Contarini, afterwards, both of them named Cardinal. Medicine was his main passion, but from humanist and scientist, he also interested to the astronomy, mathematic, physic, botany, geology, geography, and composition of verses.

He was contemporary and friend of Nicolaus Copernico, polish astronomer. In quality of astronomer, with Pietro Apiano, he affirmed that the comets' tails always appear along the direction of the sun, but in the opposite sense of it. Described also, an instrument used in astronomy, realized in the following years, by Galileo Galilei: the spyglass.

As Doctor, he's considered one of the fathers founders of the modern medicine: for the first, he hypothesized that the infections are caused by germs carriers of diseases, with the ability to multiply themselves inside the organism and to infect through the breath and different other forms of transmission.

Considering the acquired prestige in the field of the medicine, he was named, Pontifical Archiatra, and in the second time, the main Doctor of the Trent's council (1545), under the Pontificate of Paul III°.

Such charge, allowed him to be one of the craftsmen of the same council, wished from Saint Centre.

They wanted to move the council from Trent to the city of Bologna, wich was nearer to Rome and outside from the Germanic Empire.

The occasion came when, in February of 1547, he diagnosed a worrisome epidemic of typhus fever bursted in Trent.

The Papal legates, decided therefore, to transfer the Assizes to Bologna and the Pope confirmed it.

Fracastoro died the 8th August of 1553 to Incaffi, actually Affi, place near Verona his native city, where he had predominantly lived and practiced his activity of Doctor.

Cardinal Pietro Bembo (1470-1547) - Tiziano

Works:

In the 1521, Fracastoro wrote Cardinal Pietro Bembo some letters, describing an unknown illness coming from the new world and cause of epidemics, that he named for the first: Syphilis.

In August of 1530 he published in Verona an epic poem in three volumes about it. the work had 50 editions in Latin language and about 60 editions in other languages: German, French and English.

In this poem is narrated the history of Sifilo, a young shepherd who offended Apollo, God of the sun, that punished him with a terrible illness that irremediably destroys the beauty.

Really the author composed two version of the work, both of them in Latin and in prose, dedicated to his friend and University study colleague: Cardinal Pietro Bembo.

One of the two essentially literary, with the title. *"Hieronymi Fracastorii Syphilis siue morbus gallicus"* easier understanding and wide diffusion,

and the other in form of essay, entitled *"Hieronymi Fracastorii Syphilidis sive de morbo gallico"*, that results to be a scientific monograph, compiled for the Doctors; where the pathology is described in a detailed way, in its symptomatology, diagnosis and therapy.

This last version has been found in 1939 by Francesco Pellegrini and preserved in the civic library in Verona.

Fracastoro's work, was diffusedly known in Italy in its translation in vulgar Italian, *"Syphilis sive de morbo gallico"*, published by Vincenzo Bernini, in the 1765:

"....primieramente era mirabil cosa che l'introdotta infezione sovente segni non desse manifesti appieno se Quattro corsi non compia la luna...." "...tosto, pel corpo tutto, ulceri informi usciano e orribilmente il viso....".

It has not verified if Fracastoro, Doctor and Poet, created this name, Sifilo, from nothing, or extrapolated it, from others already existing.

In fact, several etymologies have been proposed by luminaries authors (Es. Sun phileo, rather, coming from the love according to Falloppia).

But everything is, the origin of this word; it's certainly possible to affirm, that the neologism had a great resonance and replaced very soon, all the other names with which, the disease was identified.

The literary choice of the poem, being praised by the contemporary, thanks to the style and the rich imagination, has been for the author, the opportunity to describe in the scientific way the Syphilis and the possible remedies with which, he believed that it can be care: mercury and guaiac, called "ligno sancto", an originates wood of the Antilles, green, with black striations, very hard and considered the heaviest wood existing in the world.

Fracastoro sustained that the syphilis (considered a divine punishment because contracted for the easy customs) could be eliminated only through a deep perspiration.

For this reason he administered the mercury, which being toxic for the salivary glands, it produced a powerful secretion that allowed the recovery.

In similar way, the "guaiac" acted, it also needs to remember, that to the tisanes and the infusions, drawn by the guaiac, the medical science, attributed, in that historical period, powers of Panacea.

In 1535, Fracastoro wrote *"De causis criticorum dierum"*, and in 1538 *"Dies critici vel de dierum criticorum causis"* texts in which he analyzed from the physician-philosophical point of view, the critical days of the illness.

It 's known also, his work *"Homocentricum"*, in which he proposed an alternative to the cosmological - tolemaico system, taking back the system of the homocentric spheres.

In 1546 the most genial work of Fracastoro was published about the medicine, entitled *"De contagione et contagiosis morbis et curatione"*.

The Italian Doctor conceived, the existence of contagion's vectors, that called "seminaria", during an epoch in which the microbes weren't known yet.

He had observed, in fact, that the illness was transmitted both for direct contact among people (contactu), that through objects, as it happens for example with the garments (fomites), both to distance as in the case of the smallpox and the plague.

In this work Gracastoro write:*" it seems that three different types of contagion exist.*

The first infects only for direct contact.

The second acts in the same way but it leaves besides tinders, and this contagion can spread out through these (tinders), as for example: the scabies, the tuberculosis, the malignant stains, and similar.

Speaking about tenders, I intend garments, wood objects and things of such sort, that although they aren't contaminated, they can preserve, however, the original germs of the contagion and infect through them.

For the third, a type of contagion exists, that transmits the illness for direct contact, through the tenders and also to distance, for example: pestilential fevers, the etisies, some types of ophthalmia , typhus and similar.

These different types of contagions, seem to obey to a determined law; rather, those that infect the far objects, do it, both for direct contact and through the tinders, those that are contagious through the tinderses are also it for direct contact, not all are contagious to distance".

Fracastoro therefore, realized the existence of micro-organisms able to transmit the infections, proposing a scientific theory on the germs, 300 years before the empirical formulation by Louis Pasteur and Robert Koch.

Nevertheless its astronomic, cosmological and cosmogonic knowledges, brought him to integrate the theory of the epidemic contagion, with the presence of the influential power of the stars, in to progress of the same epidemics.

In 1546 the work was published "*De sympathia et antipathia rerum Liber unus*", a text of natural philosophy in which Fracastoro affirms that in nature everything is connected from a natural and universal strength: it deals of the "sympathia" of everything for the part, and of the part for everything.

This strength is considered by the author, not in spiritual sense, but in physical and natural sense, according to the laws of the atomistical theory. in fact he explains, that are the flows of the atoms to establish the relationships among all the things of the world.

He sustains therefore, the attraction between the similar things and the repulsion among those different. Fracastoro refuses the explanation of the phenomenons through hidden causes, because he believes that the appeal to them, isn't an attitude that suitable to an authentic scientist.

He believes that it's always necessary, to examine deeply the facts, so it's possible to elaborate wide inductive generalizations.

Fracastoro, therefore, sustains that in all of our scientific investigations, it needs to follow to the description and the evaluation of the phenomenons.

He also wrote, three philosophical dialogues 1) *Naugerius sive de Poetica* - 2) *Turrius sive de Intellectione* - 3) *Fracastorius sive de Anima*.

These works treat the theme of the human nature, examined in its cognitive abilities and in its relationships with the cosmos and with a supernatural reality, in front of which, the speculative trial appears inadequate.

The first work, treats the poetic's theme, and it is the most meaningful of the last year of the author's life, because he strongly defends the autonomy of the art.

With an Aristotelian vision, even if, with many elements of platonic nature, the poetry is considered as universal representation, as the action through which the idea is realized in its visible beauty.

It's recognized besides, the freedom of the artist, to transpose the universal beauty, inside to the objects, in sensitive forms.

In the "*Turrius sive de Intellectione*", a dialogue to " gnoseologic"(gnosis γνοσις- logos λογος) character, the author faces the intellect's theme, considering the logic as the tool of natural knowledge. The tradition of the philosophers of end 300 and 400, which had elaborated deep analyses about the different forms of the sensitive experience, it's taken back and developed, in particular way, about the perspective and the optics. The last work of the author is "*Fracastorius sive de Anima*", a dialogue of psychological matter. In this text the thought of Fracastoro manifests it, with a series of questions that the author do to himself, at the end of his experience of life, as doctor and philosopher. The work results to be incomplete.

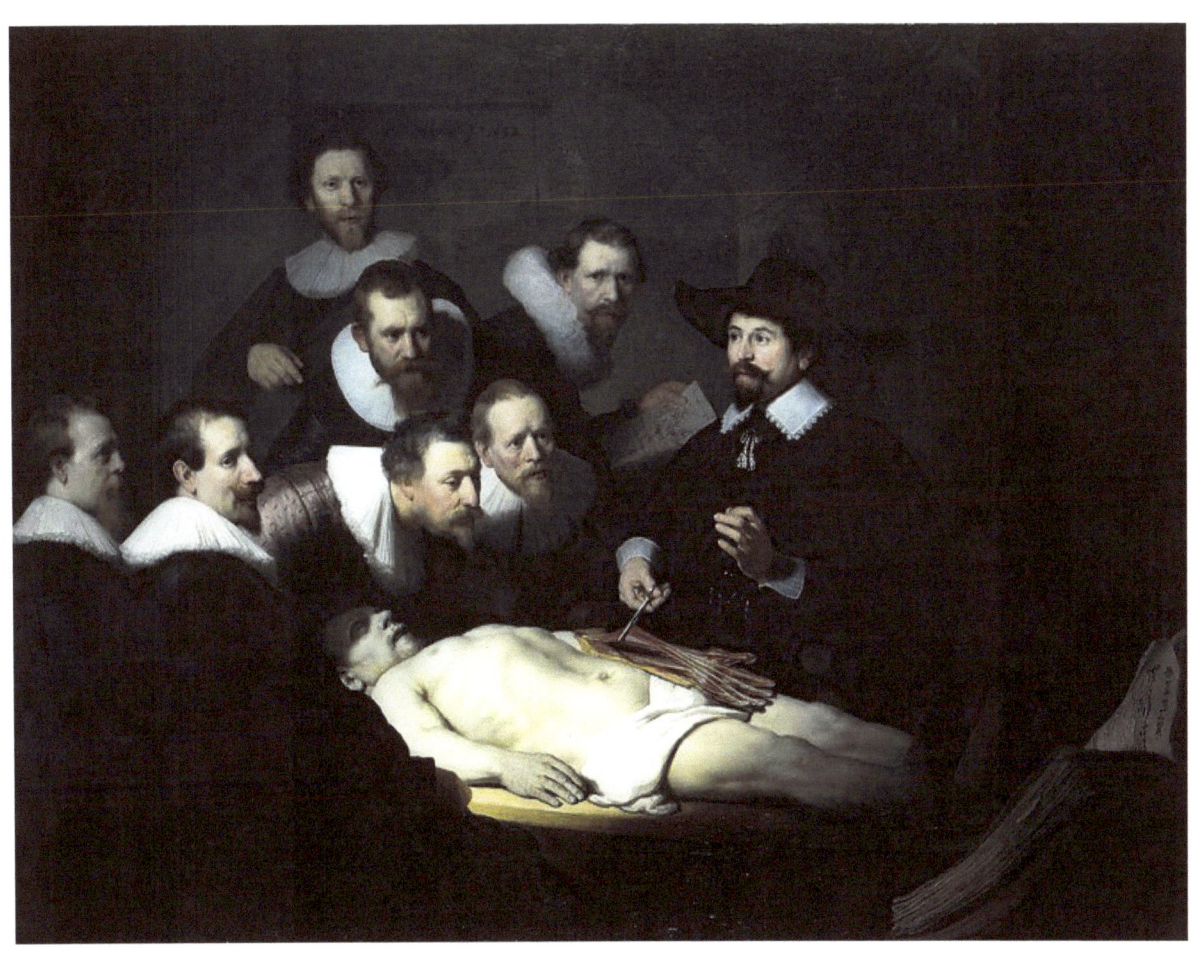

The Anatomy Lesson of Dr. Nicolae
Rembrandt (1606-1669)

Five centuries after the epidemic of syphilis, another venereal disease is spreading, finding the current medicine completely unprepared, it has made to speak again the Church of sexual abstinence, the sanctimonious people speak of divine punishment,
what the men of the 20th century have called AIDS.

BIBLIOGRAPHY

Apolant: "Ehlrich", Berlin, 1914.

Astruc: "De morbis November venereis books", Paris, 1740.

Arrizabalaga J., Sebastian Aquila (c. 1440 - c. 1510), and "ill francés" y la "dispute of Ferrara" (1497),
in "Acta Hispanica to Medicinae Scientiarumque
Historiam Illustrandam, " vol. XIV, 1993, pp.. 227-247.

Barbarani: "Fracastoro and his works", Zannoni, Verona, 1897.

Bellini: "History of Dermatology", Milan, 1934.

Bethengourt de Jacques, "Nova penitentialis Lent,
nec not purgatorium in morbum gallicum, seu venereum;
Cumulative a silver box aque et Ligny guaiaci
colluctantium super dicti diseases curationis prelatur, opus
fructiferum; Parisiis, 1527.

Brown: "Arabian Medicine", Cambridge, 1923.

Campbell: "Arabian medicine," London, 1926.

Canevari D.: "De ligno St. commentarium" Romae,
Apud Gulielmum Facciottum, 1602.

Capparoni: "A description of syphilis made by
Bartolomeo della Rocca said Cocles, published in Bologna
in 1504, "Acts and Memory Accad. St. Art Sanit., 1935.

Castiglioni Arturo: "Girolamo Fracastoro and his Poem
De Morbo Gallico, "Rass. Clin. Scient. Dept. Bioch. Ital,
1927.

Cipriani M., "Contribution to the study of etymology word syphilis ", Science Magazine St. Nat Med., 1948.

Delgado Francois: "On the manner of using the wood saint or the way that you heal the French disease, and every ill incurable "Venise, 1529.

G. Del Guerra: "Venereal Prophylaxis and songs of the people cheering, " Bull Art Institute St. Sanit., Rome, 1930.

G. Del Guerra: "The descent of Charles VIII and the clinical diary of lue by John Mariani (1496-1497) ", Giardini, Pisa, 1955.

C. Di Cicco: "History of Syphilis" 14th Congress of European Academy of Dermatology & Venereology, London. JEADV Vol.19, Suppl.2, 1 - 411, 2005.

Girolamo Fracastoro's "Syphilis sive de Morbo Gallico", books tres to Petrum Bembum, Patavii, J. Cominus, Editio II, 1739.

Girolamo Fracastoro: "Opera Omnia" editio secunda,Venetiis, apud Juntas, 1754.

Fracastoro, Girolamo [1478-1553]. Hieronymi Fracastorii Syphilidis: sive de Morbo Gallico = introduction, version and notes by Francis Pellegrini. - [Verona] Verona editions of Life, 1956 (CIVVR, SLVR) .

Fracastoro, Girolamo [1478-1553]. Hieronymi Fracastorii Syphilidis, sive morbi gallici lib. III. Joseph lib. II. item Carminum lib. I. Rutilii Claudians Numatiani Galli u.c. itineraria. - Antuerpiae: Apud Martini Nutij Viduam, 1562 (CIVVR).

Fracastoro, Girolamo [1478-1553]. Hieronymi Fracastorii Syphilis, sive morbus gallicus. - Veronae: [S. Nicolini and f.lli], 1530 (CIVVR).

Girolamo Fracastoro: "On the syphilis or de morbo Gallico", 3 books, vulgarized by Vincenzo Benini Colognese Della Volpe, Bologna, 1765.

Gangolphe: "The syphilitic lesions prehistoric", Mem Acad. Science. Lyon, 1912.

Garrison: "Fracastoro," Science, New York, 1910.

Giuliano Gliozzi: "The discovery of the savages. Anthropology and colonialism from Colombo to Diderot, Milan, Monaco, 1971.

Gruzinski Serge: "The colonization of the imagination. Indigenous societies and westernized in Mexico Spanish. "Einaudi, Torino, 1994.

Laita Pierluigi. "The salary of Girolamo Fracastoro. Medical Council of Trent, 1545-1547. "Verona, 1971 (CIVVR).

Serge Latouche: "The Westernization of the world", Turin, Basic Books, 1992.

Konrad Schellig "In pustolas malas morbum quem malum de que sunt de France vulgus appellat generally formicarum consilium healthy ", Heidelberg 1496.

Malamani Anita: The Hospital of the Incurables of Pavia origins to its uptake in P. L. Pertusati (1556-1796)

Oviedo: De la natural historia de las Indias, or de Sumario the natural, etc.. printed in 1526 Toledo.

Barcia de las Indias in Historiadores occidentales - Volume I - Madrid 1749

Poll Nicolas: "gall diseases to cure De lignum guayacanum libellus "Venetiis, 1535 - Basel, 1536

Torrella Gaspar. "Tractatus contra cum consiliis

pudendagram seu morbum gallicum "Pietro Della Torre Rome, 1497.

Johann Widmann "Tractatus de Parkinson et pustulis here vulgato appointments malady Franzos appellatur ", Rome, Stephan Plannck, 1497.

Pennacchia T.: "History of Syphilis", Giardini, Pisa, 1961.

Simonelli: "Around the testimony of Oviedo and Diaz" Sexual Studies, 1924.

Sudhoff: "Mal Franzoso in Italien", Giessen, 1912.

Kampmeier, R. H. (1972). "The Tuskegee Study of untreated syphilis. " South Med J 65 (10): 1247-51. PMID 5074095.

Kampmeier, R. H. (1974). "Final Report on the Tuskegee syphilis study." South Med J 67 (11): 1349-53. PMID 4610772.

Brandt AM. Hastings Cent Rep. 1978 Dec; 8 (6) :21-9. "Racism and research: the case of the Tuskegee Syphilis Study ". PMID: 721302 [PubMed - indexed for MEDLINE]

Ralph V. Katz, S. Steven. Kegeles, Nancy R. Kressin, B. Lee Green, Min Qi. Wang, A. Sherman James, Stefanie Luise. Russell, Cristina Claudio:

"The Tuskegee Legacy Project: Willingness of Minorities to Participate in Biomedical Research ".

Journal of Health Care for the Poor and underserved Volume 17, Number 4, November 2006 pp. 698-715.

Harriet A. Washington, Medical Apartheid: The Dark History of Medical Experimentation on Black Americans from Colonial Times to the Present, 2007. Amazon Kindle.

Ralph V. Katz, B. Lee Green, Nancy R. Kressin, S. Stephen Kegeles, Min Qi Wang, A. Sherman James, Stefanie L.

Russell, Cristina Claudio, and Jan M. McCallum. "The Legacy of the Tuskegee Syphilis Study: Assessing the ITS Impact on Willingness to Participate in Biomedical Studies ".
Journal of Health Care for the Poor and underserved, Volume 19, Number 4, November 2008, p. 1168-1180.
Chris Mc greal "U.S. says sorry for" outrageous and Abhorrent "Guatemalan syphilis tests." The Guardian, 1 October 2010.
Donald G. McNeil, Jr. "U. S. Apologizes for Syphilis Tests in Guatemala". The New York Times. 1 October 2010.
The Guardian: "Shocking new details of U.S. STD experiments in Guatemala. " Associated Press guardian co.uk. Tuesday 30 August 2011 08:12 BST .

INDEX